What is the Sun?

The Sun is a **star**. Stars are huge balls of hot gases spinning in space. Our Sun is larger than one million Earths. At its core the temperature may reach 15 million degrees centigrade. The Sun is so big and so hot, it gives heat and light to the Earth, about 150 million kilometres away. YOU MUST NEVER LOOK DIRECTLY AT THE SUN. IT IS **SO** BRIGHT IT WILL DAMAGE YOUR EYES.

▽ Light and heat from the Sun reach the Earth.

Day and night

Stars shine all the time. Our Sun is so bright, it hides the light of other stars. We have day and night because the Earth spins, or **rotates**, on its **axis**. The axis is an imaginary line going through the centre of the Earth. It takes 24 hours for the Earth to spin round once. As it rotates, part faces away from the Sun. That part is in darkness, so it is night-time.

▷ Stars at night shine with a steady light. They seem to twinkle because their light is broken up as it passes through the Earth's atmosphere.

▽ As one side of the Earth moves away from the Sun, the Sun seems to sink below the horizon, filling the sky with rich, dark colours.

The Sun and Stars

Lesley Sims

W
FRANKLIN WATTS
LONDON•SYDNEY

This edition 2001

Franklin Watts
96 Leonard Street
London EC2A 4XD

Franklin Watts Australia
56 O'Riordan Street
Alexandria, Sydney
NSW 2015

ISBN 0 7496 4138 X

A CIP catalogue record for this book is available from the British Library

10 9 8 7 6 5 4 3 2 1

Printed in Italy

Contents

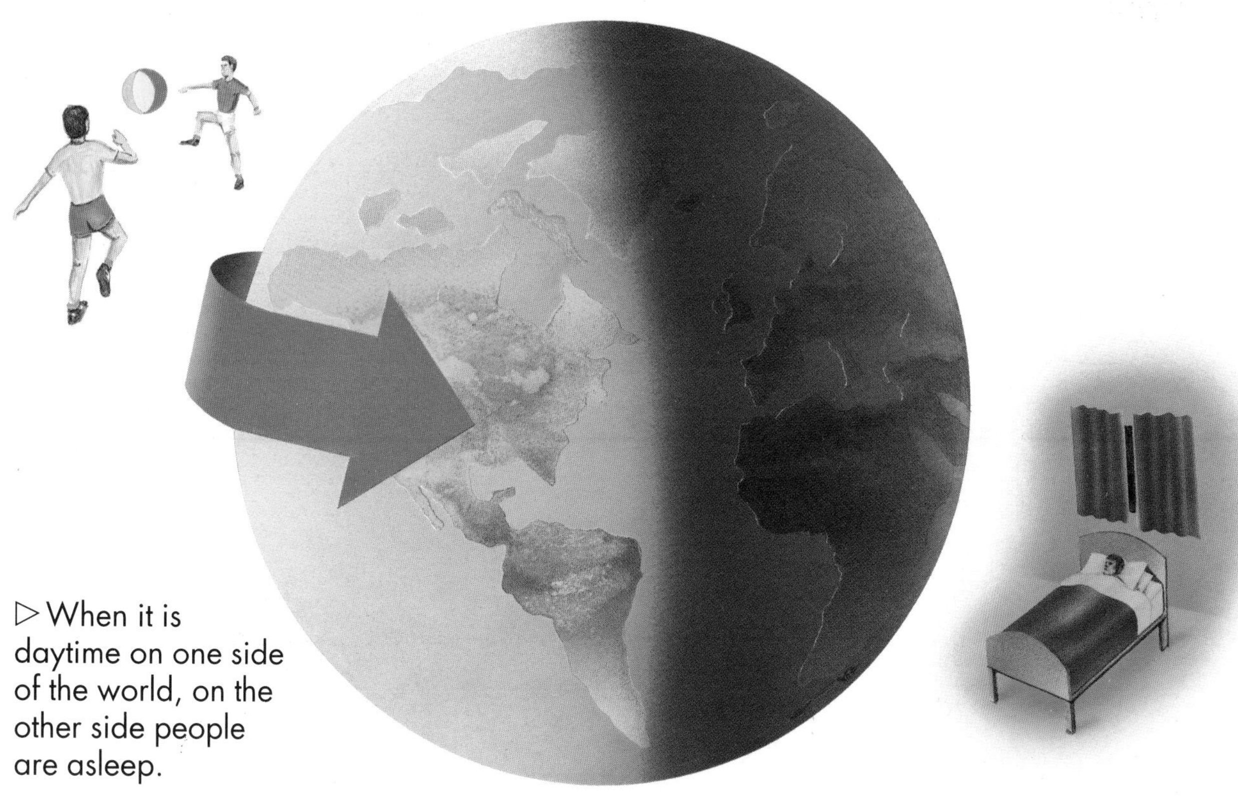

▷ When it is daytime on one side of the world, on the other side people are asleep.

Light and dark

▷ The shadow on this sundial is pointing to 5 o'clock.

As the Earth rotates, the Sun seems to move across the sky. In fact, it is the Earth which is moving. When we see the Sun, we see it from different positions. If an object blocks the light of the Sun, a dark patch, or shadow, is formed behind it, in the shape of the object. Years ago, people told the time from the changing positions of shadows.

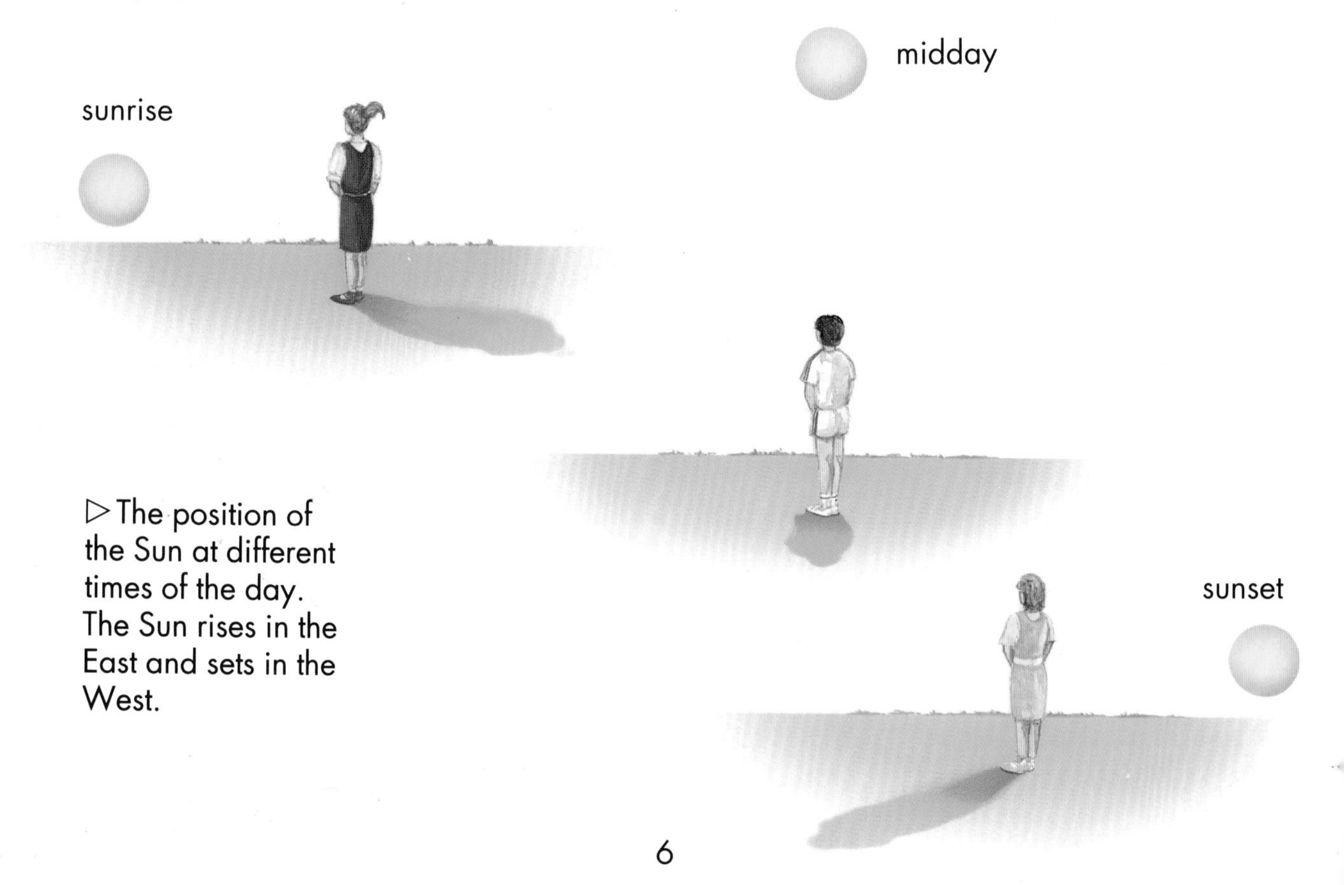

▷ The position of the Sun at different times of the day. The Sun rises in the East and sets in the West.

▽ An object which blocks the Sun is opaque. You cannot see through it.

△ An object which lets sunlight through is transparent.

On the Sun

Sometimes the Sun has freckles! They are called sunspots. They are darker patches which are cooler than the rest of the Sun. The spots appear to travel across the Sun. This is because the Sun is not still but rotates. Above the sunspots, streams of gas called solar flares shoot out from the Sun. Sometimes giant clouds of gas erupt on the surface.

▽ A solar storm which burst out of the Sun in December 1973.

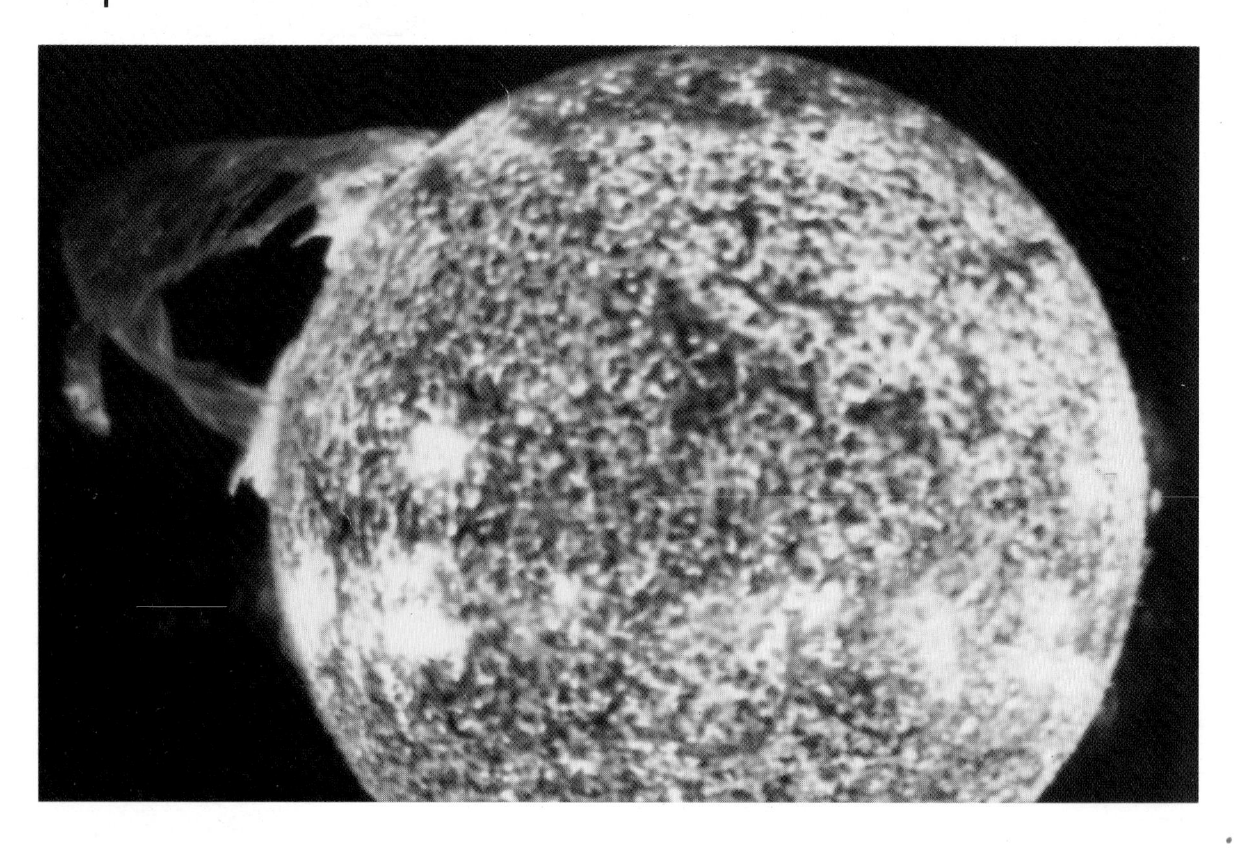

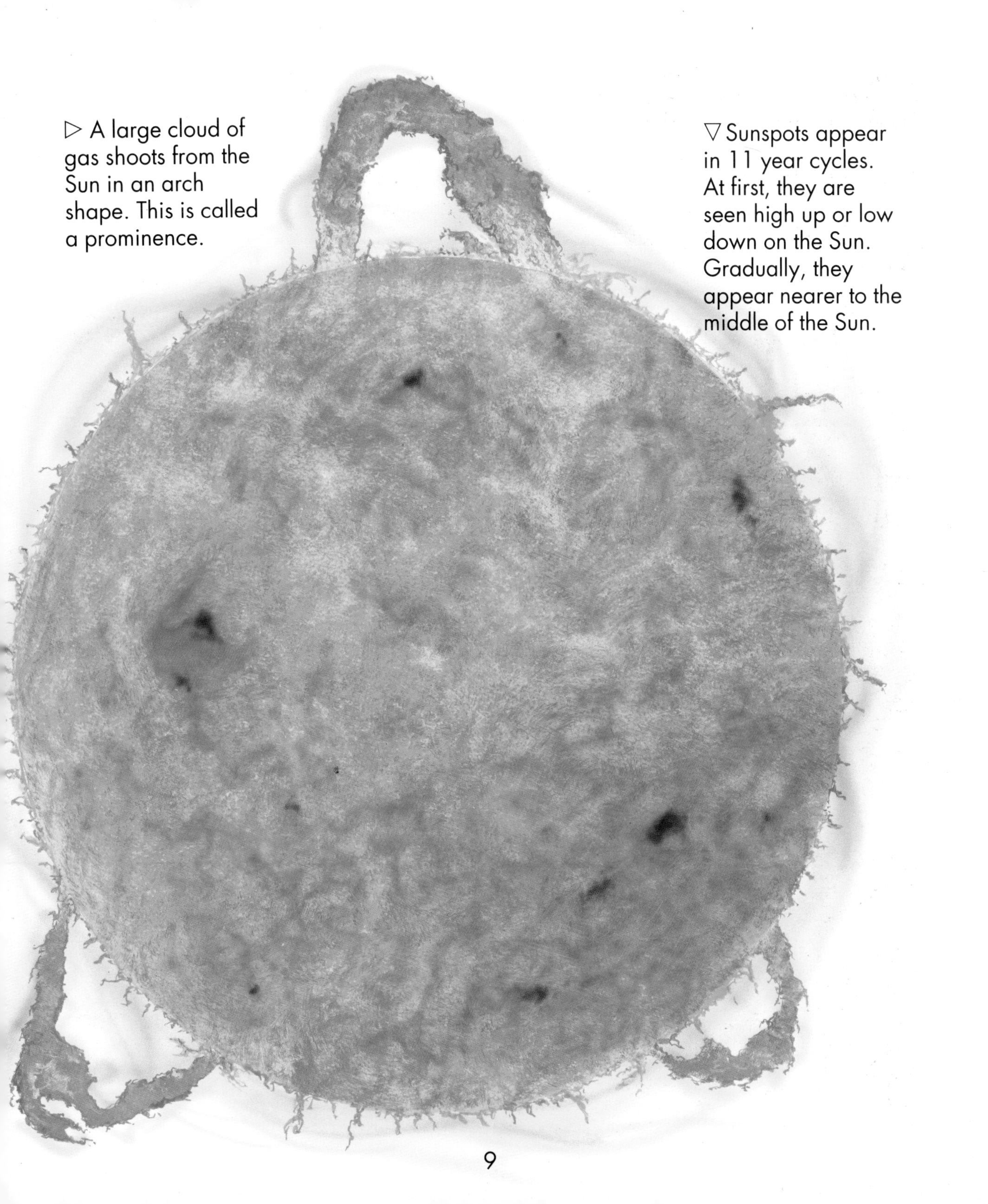

▷ A large cloud of gas shoots from the Sun in an arch shape. This is called a prominence.

▽ Sunspots appear in 11 year cycles. At first, they are seen high up or low down on the Sun. Gradually, they appear nearer to the middle of the Sun.

An eclipse of the Sun

The Sun is much bigger than the Moon, but it is much further away, so they look the same size. As the Earth **orbits** the Sun, the Moon is orbiting the Earth. Sometimes the Moon is exactly between the Earth and the Sun. It blocks the light of the Sun, casting a shadow over part of the Earth. This is an eclipse of the Sun. During an eclipse the Sun's **corona** can be seen.

▽ When the orbit of the Moon brings it into a direct line between the Sun and the Earth, there is a total eclipse.

▽ If the Moon only covers part of the Sun, there is a partial eclipse.

◁ At the end of a total eclipse, the Sun begins to peep out from behind the Moon. It looks like a diamond ring.

▽ The corona, photographed during an eclipse. The Sun is so bright, it normally hides this outer layer of gas.

Just a star

The Sun is not a special star. Other stars are hotter, or older, or larger. The Sun seems so huge to us because it is the closest star to the Earth. The next nearest star is 40 million million kilometres away. The distances between stars are so large that they are measured in **light years**. This is the distance travelled by light in one year.

▷ Proxima Centauri, one of the stars in this large, bright group, is our nearest star after the Sun. It is 4.3 light years away. So it takes over four years for light from this star to reach the Earth.

△ The Sun is only an ordinary star but it is essential for life on Earth. Without the Sun these crops could not grow.

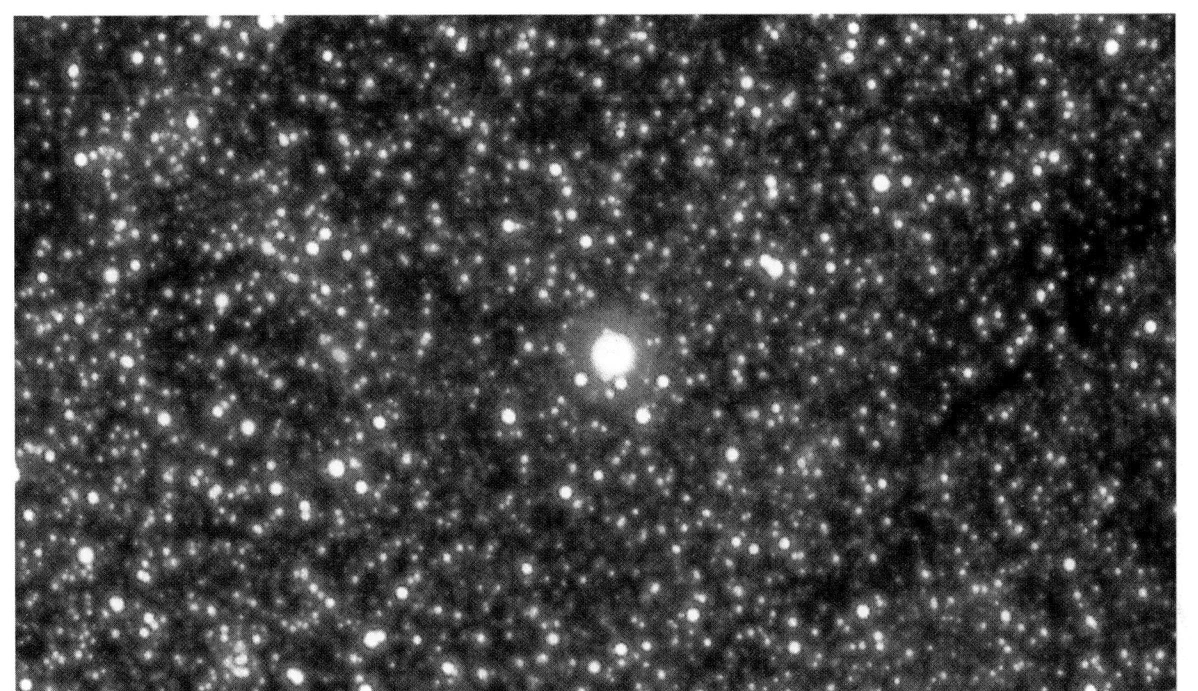

▷ As a star grows older, it gets bigger and cooler. It gets redder in colour.

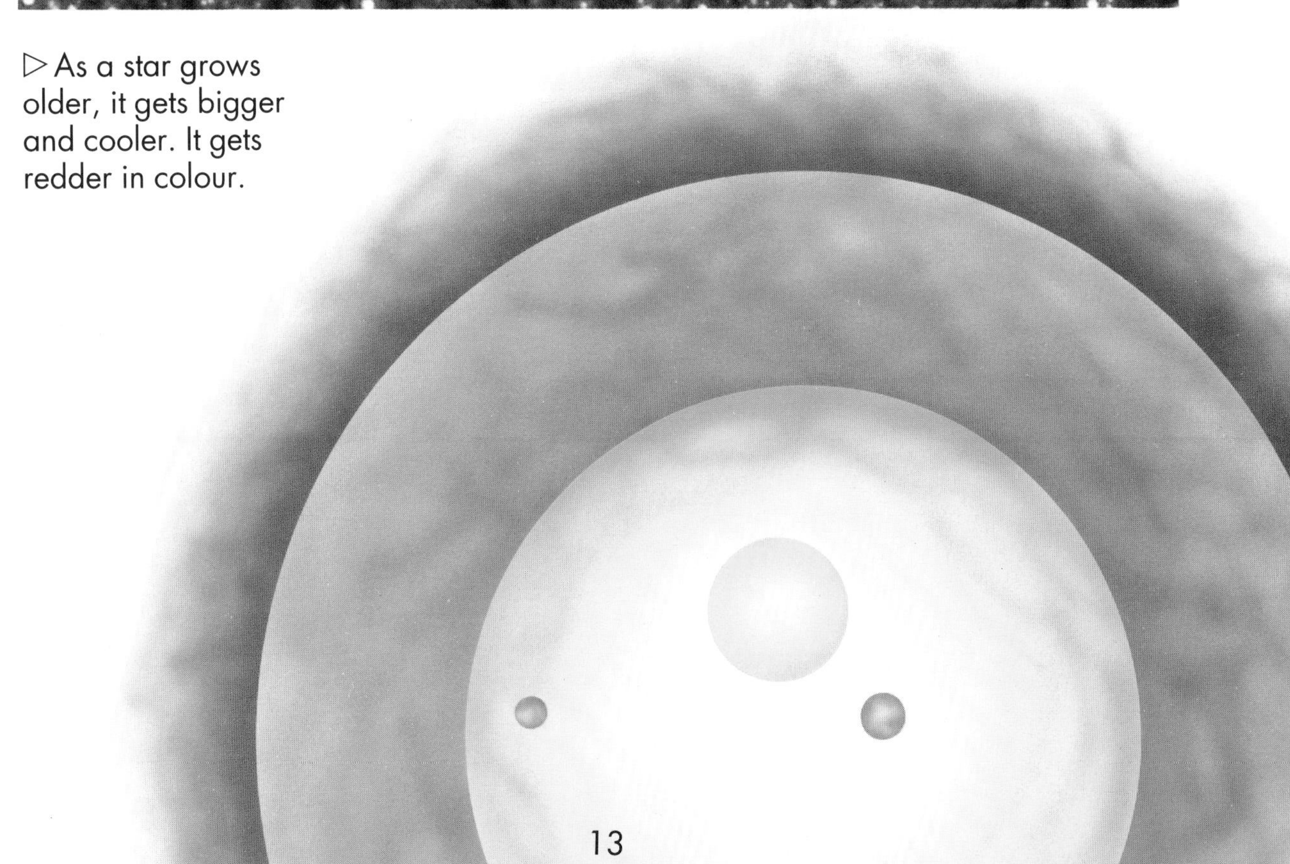

The life of a star

▷ Betelgeuse, a red giant.

Stars form out of vast clouds of gas and dust. They shine for millions of years. As a star like the Sun grows older, it uses up its gas. It grows enormous and becomes a red giant. Then its outer layers fall away into space. The star shrinks into a white dwarf. Finally, with no energy left, it shrinks into a black dwarf, becoming a cold, dark lump.

▽ The life of a star like the Sun goes through several stages.

A star forms in a gas cloud, or nebula.

It burns its gas for many millions of years.

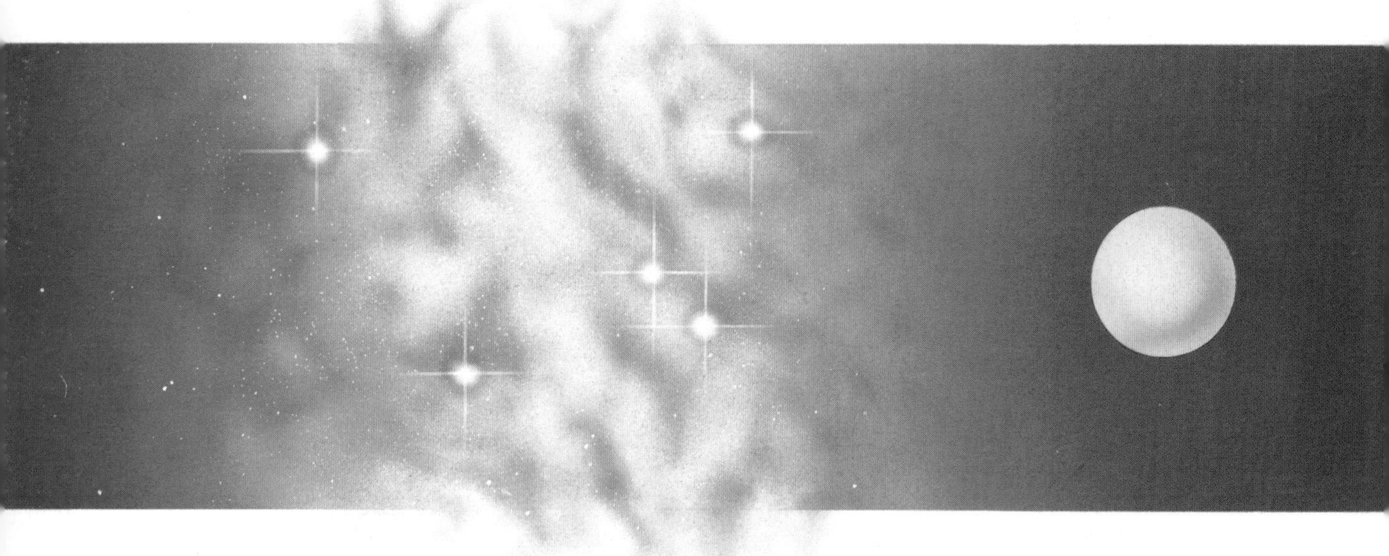

It expands into a red giant.

It shrinks into a white dwarf.

A black dwarf.

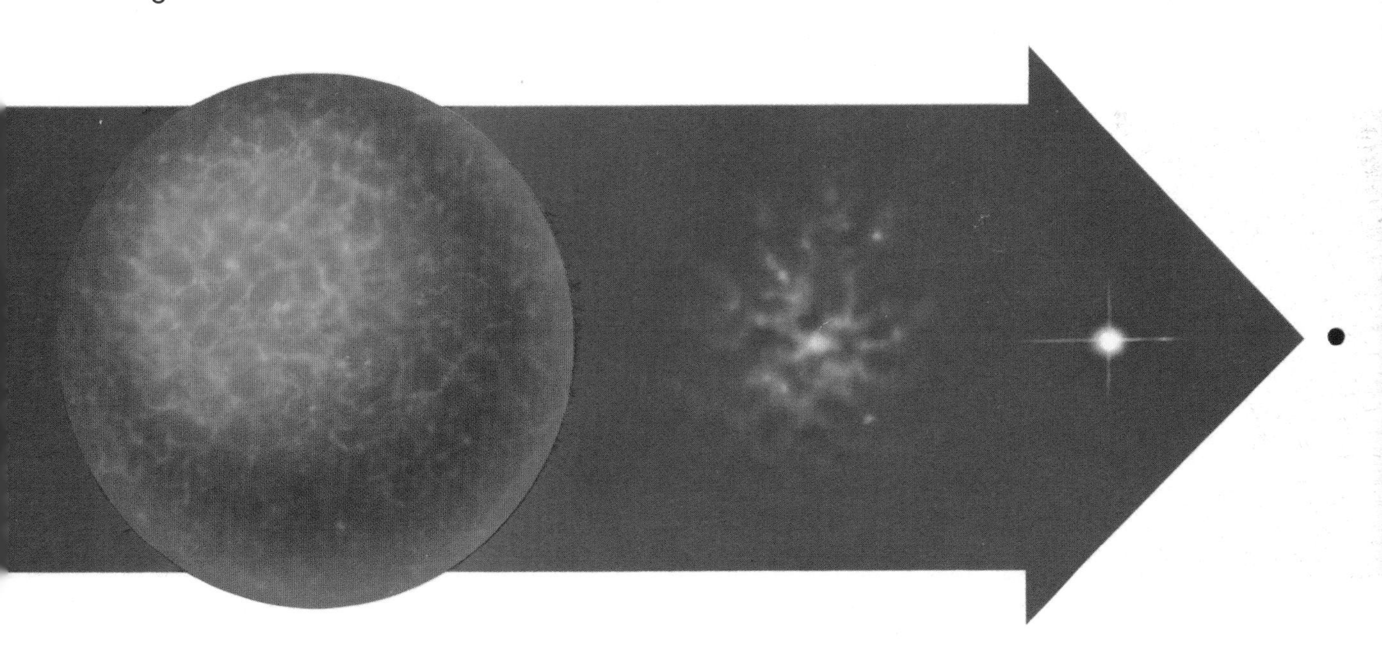

Supernovae

Some stars are so large they do not simply fade away into white dwarfs. These are stars that are more than three times the **mass** of the Sun. They expand into supergiants as their gas is used up. Then they explode and become supernovae. The explosion is brighter than the light from millions of suns. A tiny neutron star, or pulsar, is left.

▷ This cloud of gas is the remains of a supernova which exploded in 1054, over 900 years ago.

▽ A supernova is so violent, some scientists believe it starts new stars forming.

A bright star with a large mass.

The star expands and becomes a supergiant.

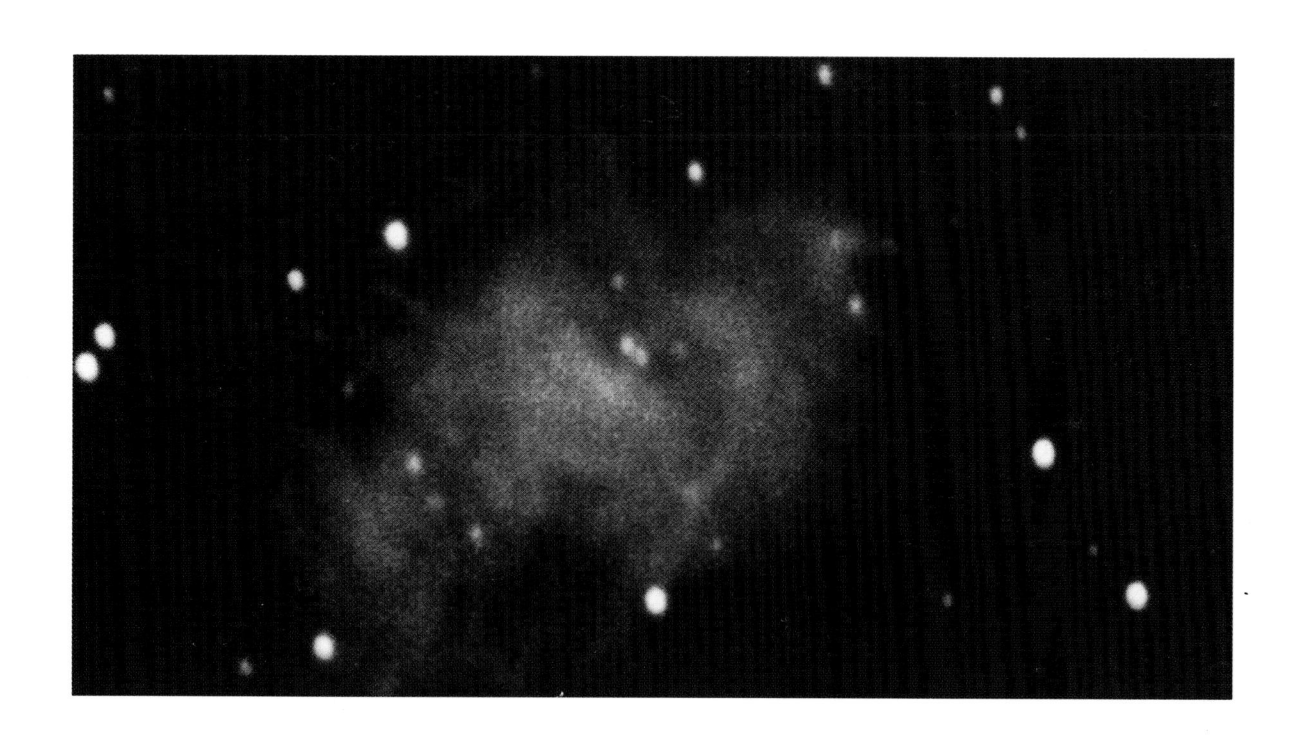

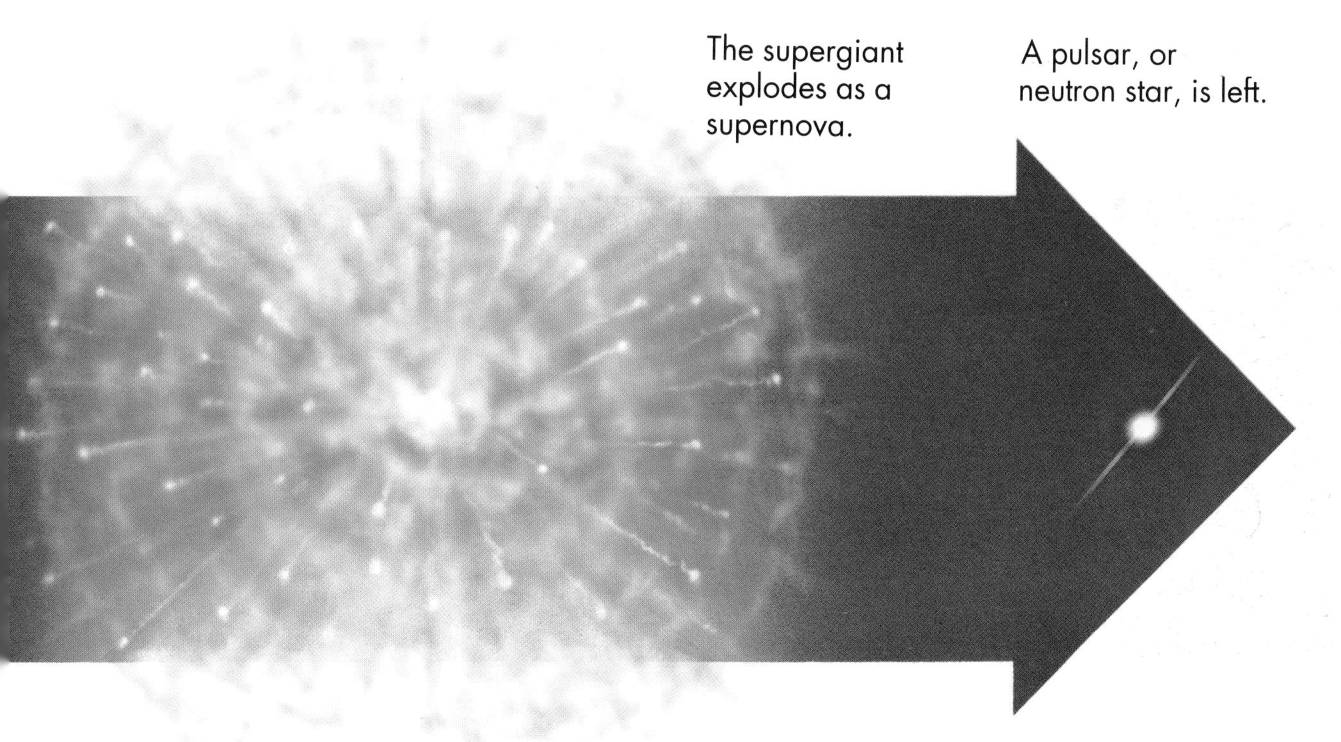

The supergiant explodes as a supernova.

A pulsar, or neutron star, is left.

Starlight

Light is made up of different colours. The light of a star can be split into colours. The colours in light are called the **spectrum**. Scientists can find out what a star is made of by studying its spectrum. When sunlight shines through raindrops, it is split into the colours of the spectrum. A rainbow appears in the sky.

▷ The water droplets splashing from the waterfall split the light of the Sun.

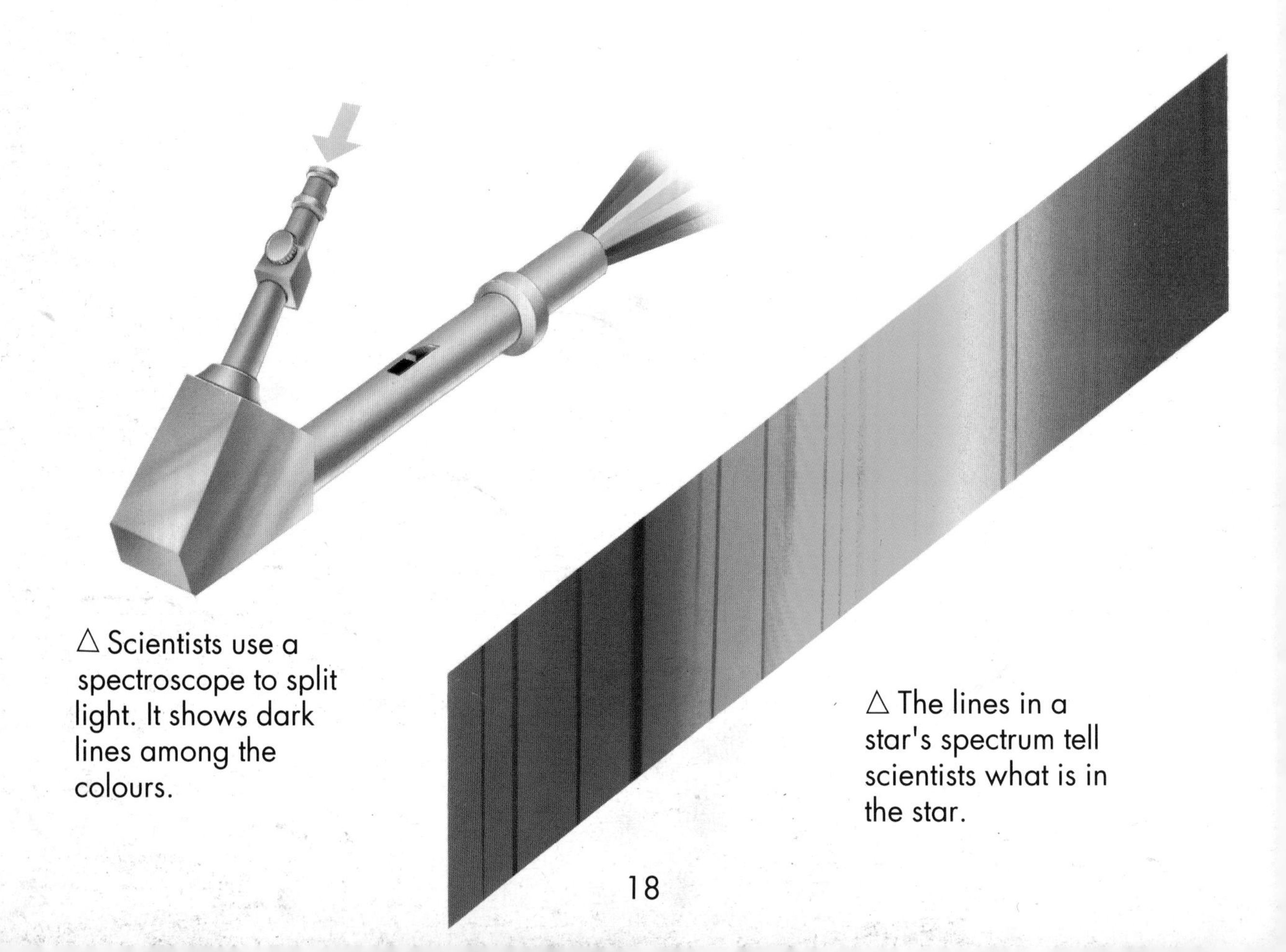

△ Scientists use a spectroscope to split light. It shows dark lines among the colours.

△ The lines in a star's spectrum tell scientists what is in the star.

Galaxies

Groups of millions of stars are called **galaxies**. Galaxies have different shapes. They are named after the shapes they make in the sky. We live in a spiral galaxy. It looks like a spinning catherine wheel, with arms spiralling round. Some galaxies are round or oval. Some are made up of millions of stars jumbled together. These are irregular galaxies.

▽ The universe is made up of thousands of millions of galaxies with empty space between them.

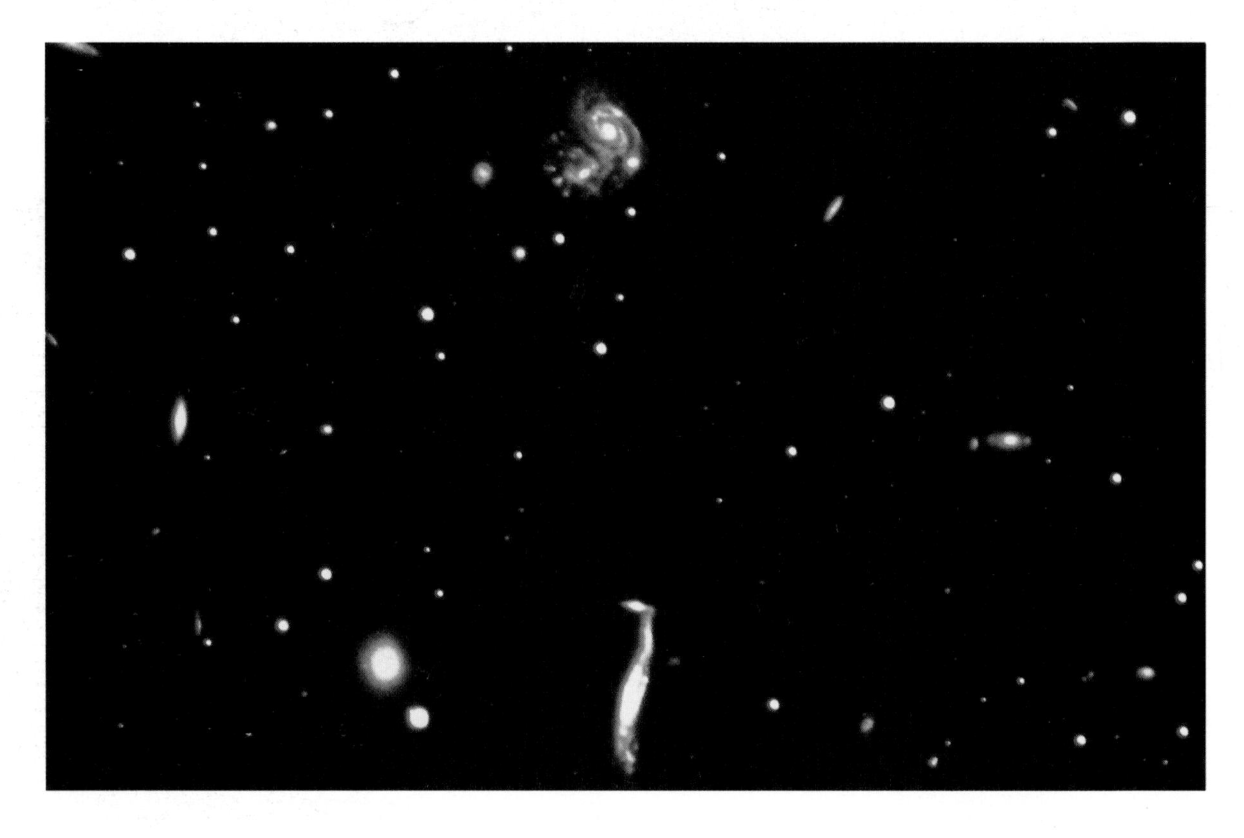

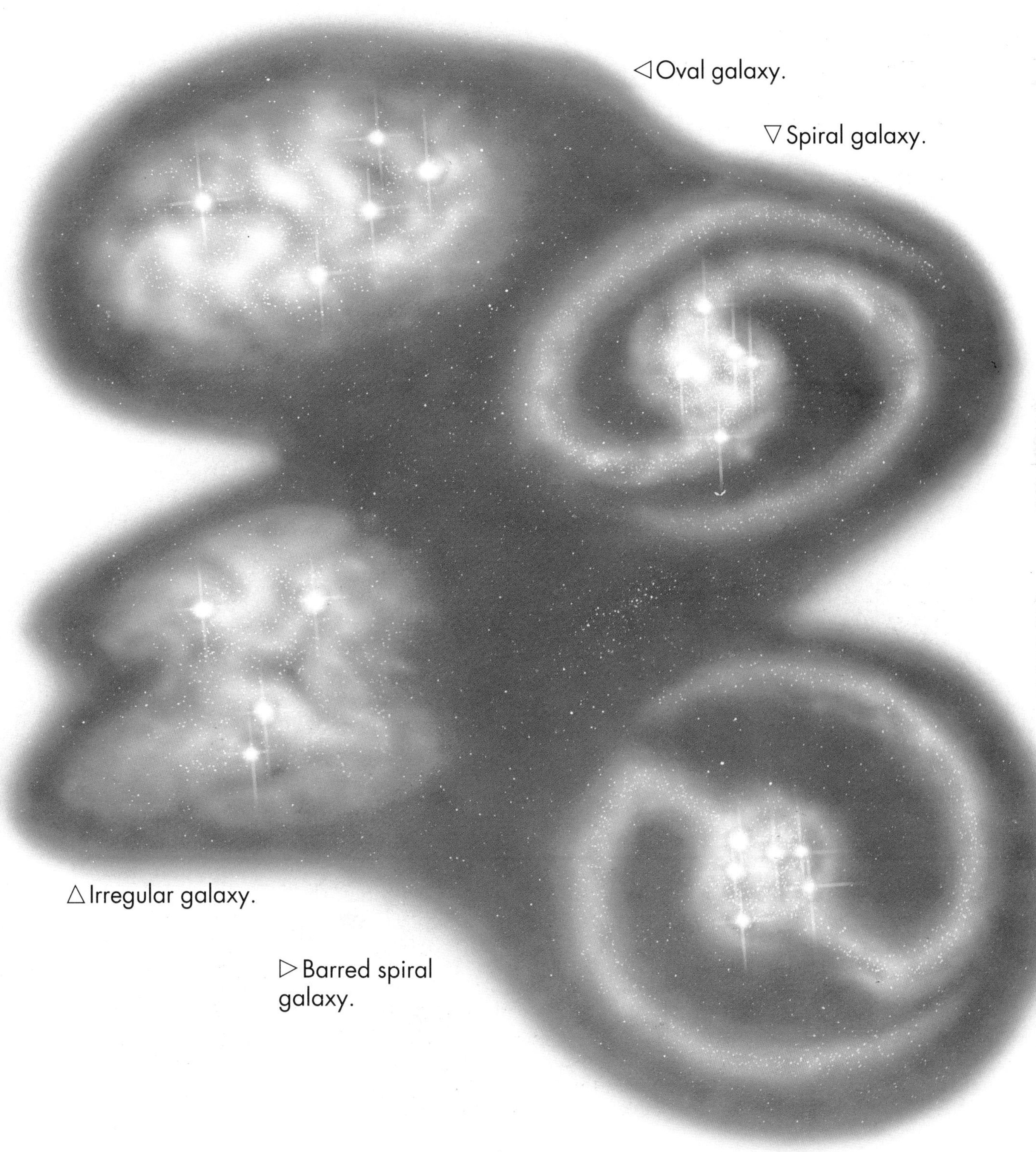

◁ Oval galaxy.

▽ Spiral galaxy.

△ Irregular galaxy.

▷ Barred spiral galaxy.

Where do stars come from?

The universe was born about 15,000 million years ago. Scientists believe that before then everything was squashed together in a lump. Suddenly there was a huge explosion. Pieces of the lump were thrown out, in a whirling cloud of gas and dust. The cloud probably separated into blobs which formed galaxies. Later, smaller blobs inside the galaxies became stars.

▽ Our Galaxy is the Milky Way Galaxy. The part we can see from the Earth is known as the Milky Way. The stars make a creamy white band across the sky.

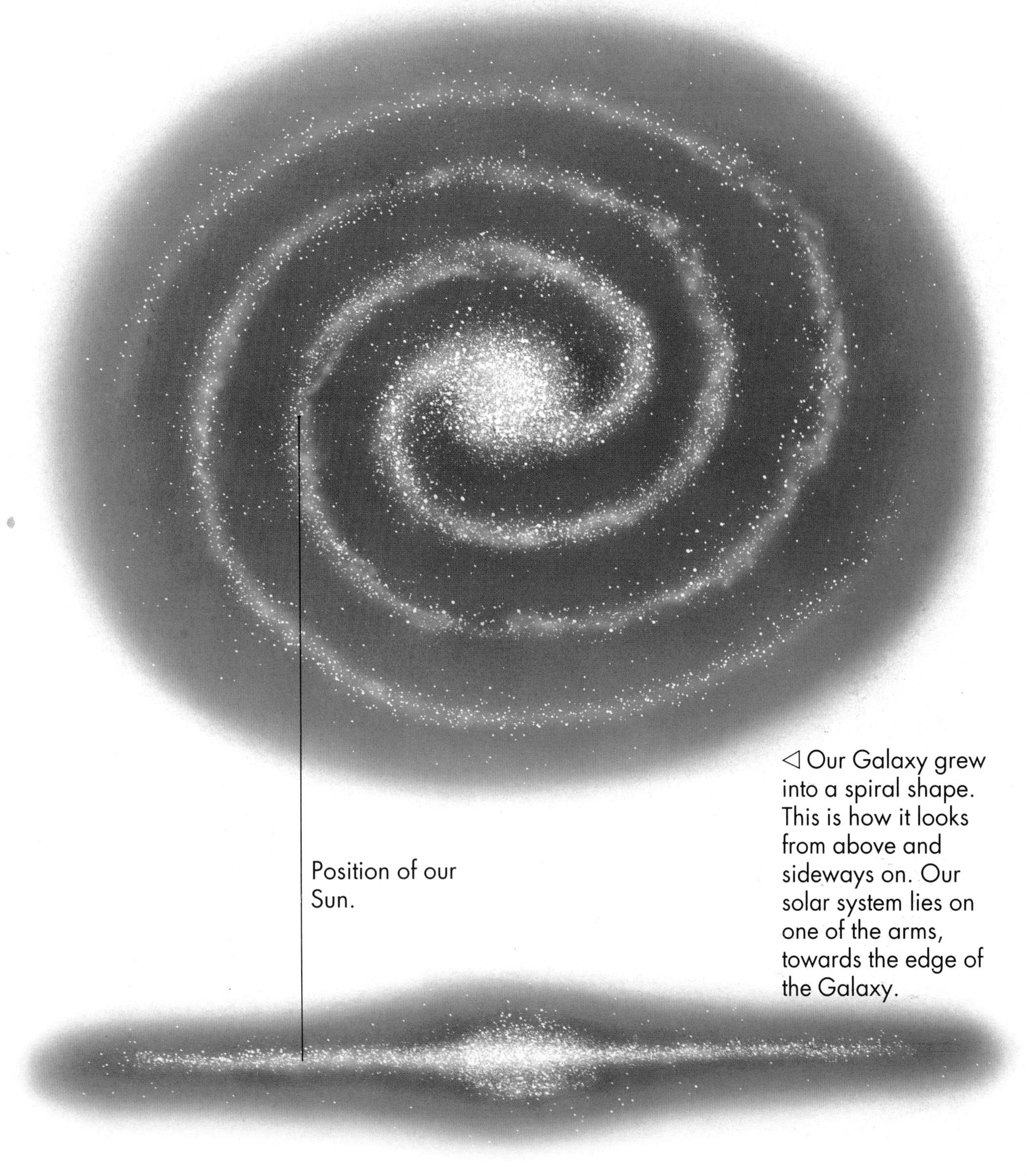

◁ Our Galaxy grew into a spiral shape. This is how it looks from above and sideways on. Our solar system lies on one of the arms, towards the edge of the Galaxy.

Star clusters and nebulae

Star clusters are groups of stars within galaxies. There are two kinds of star cluster. Open clusters are loose groups of hundreds of young stars. Globular clusters are made of millions of stars packed tightly together, in a ball shape. **Nebulae** are clouds of gas and dust. Some have their own light. Others are lit up by stars. Some are dark.

▷ A globular cluster. In our Galaxy, there are over 100 known globular clusters.

▽ The Horsehead Nebula resembles a horse looking back at the stars.

△ Inside the Orion Nebula, new stars are forming.

Constellations

Constellations are patterns of stars. The Ancient Greeks saw pictures in groups of stars. They named them after animals and the heroes of their legends. Today the sky is divided into 88 constellations.

Sirius is the brightest night-time star. It is sometimes called the Dog Star, because it is in the constellation of the Large Dog.

▽ The constellation of Orion, the Hunter. A sword of stars hangs from his belt.

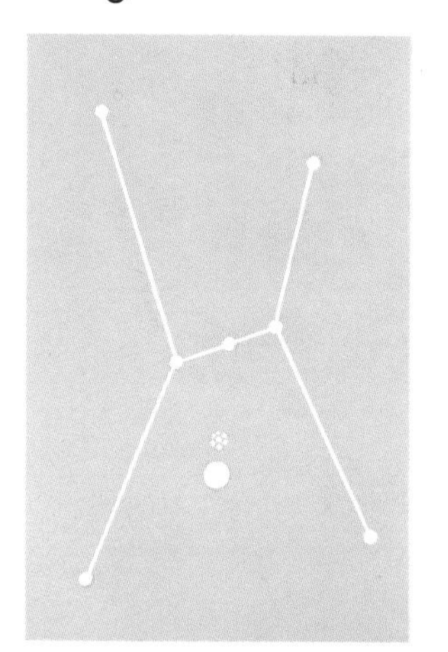

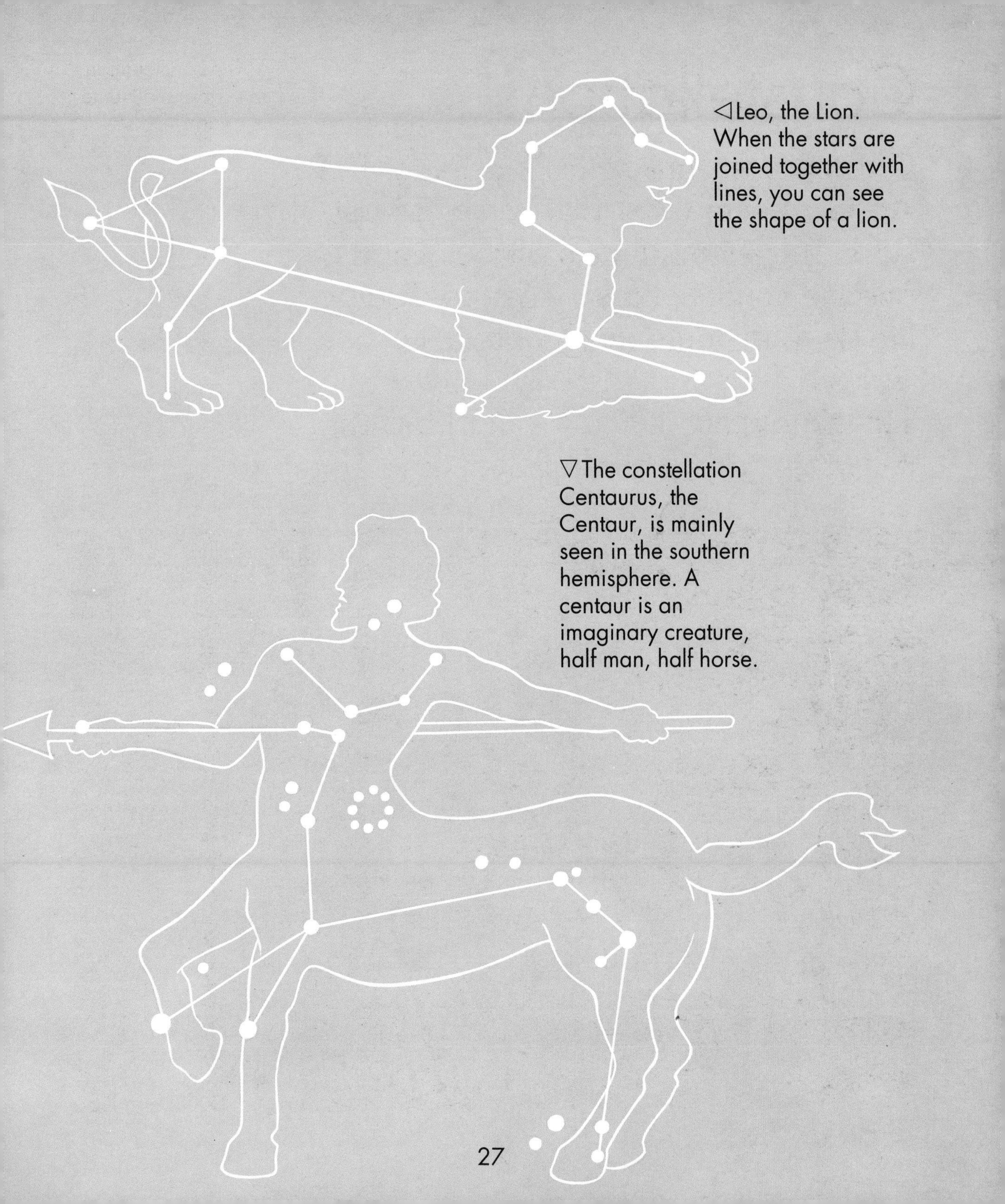

◁ Leo, the Lion. When the stars are joined together with lines, you can see the shape of a lion.

▽ The constellation Centaurus, the Centaur, is mainly seen in the southern hemisphere. A centaur is an imaginary creature, half man, half horse.

Star gazing

Two well-known constellations in the **northern hemisphere** are Ursa Major and Ursa Minor, the Great Bear and the Little Bear. In the Great Bear, a group of seven stars forms the Plough or Big Dipper. Two stars at the end of the Plough point to the Pole Star, which shines above the North Pole. The Pole Star is the brightest star in the Little Bear.

▷ The Great Bear. Can you find the Plough?

▽ Because the Earth rotates, the stars seem to move across the sky. The Pole Star is the only star which seems to stay in the same place.

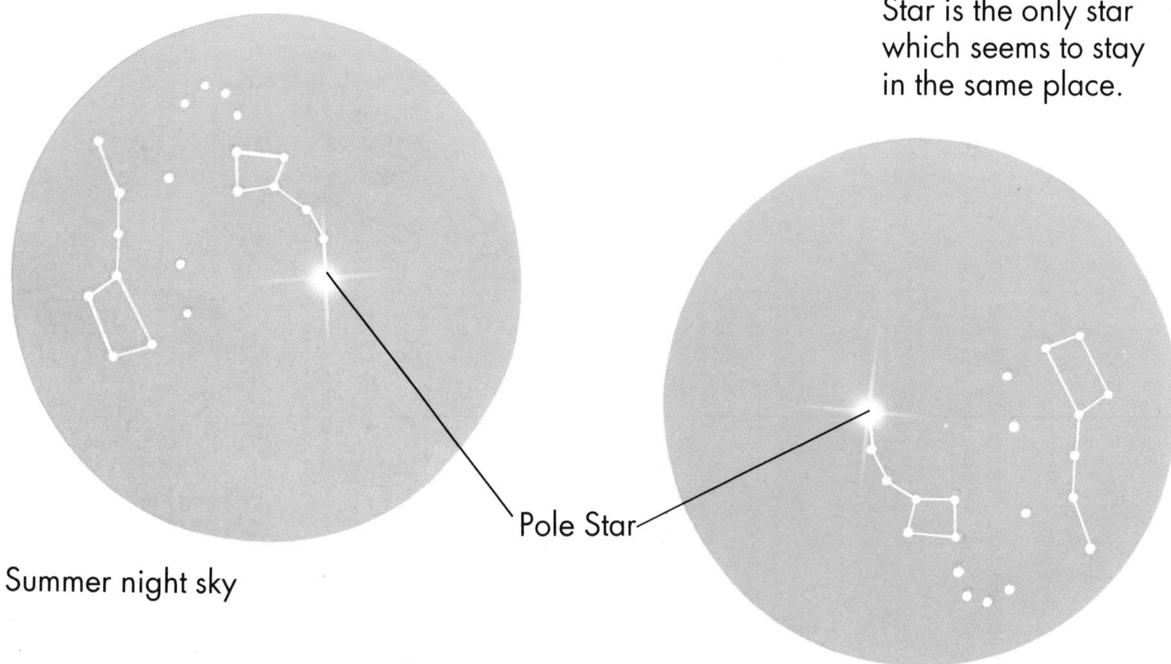

Summer night sky

Winter night sky

▽ The Southern Cross is near to the point above the South Pole. But there is no South Pole star.

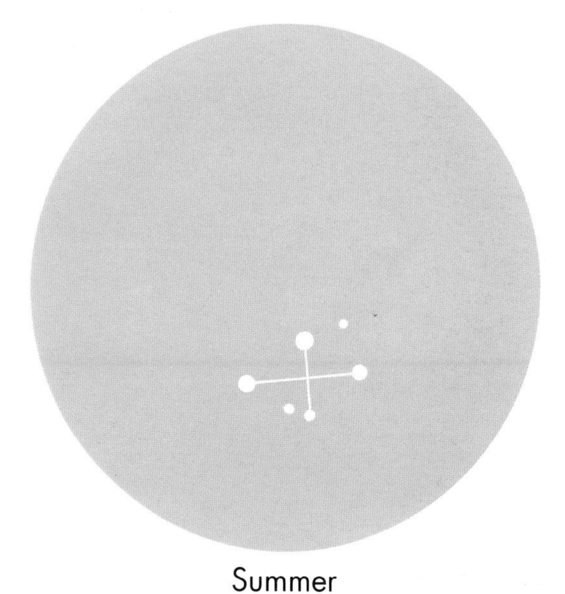

Summer

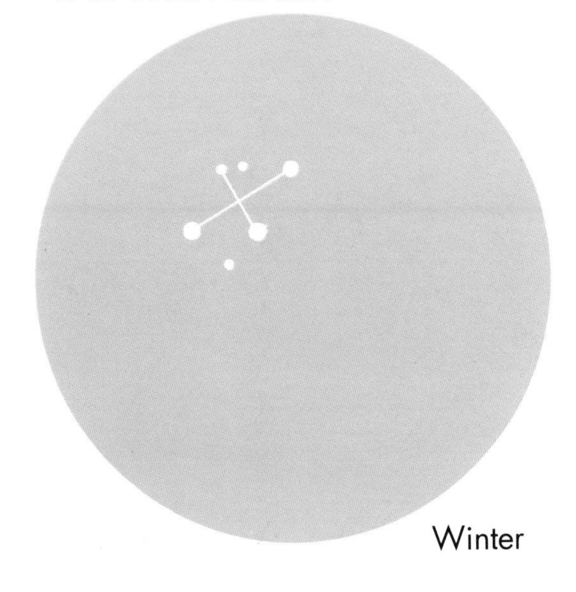

Winter

Things to do

- Shine a torch on a spinning globe to see day and night.

- Make a spinning spectrum circle: divide into six segments, coloured red, orange, yellow, green, blue and purple. When spun fast the circle appears to be white.

- Go outside (with an adult) and look at the stars. Can you find some of the constellations in this book?

Useful addresses:

Junior Astronomical Society,
36 Fairway,
Keyworth,
Nottingham NG12 5DU

British Astronomical Association,
Burlington House,
Piccadilly,
London W1V 0NL

Glossary

atmosphere The layer of gases which surrounds some planets. On Earth it allows us to breathe.

axis An imaginary line which runs through the centre of the Earth and around which the Earth spins.

constellation A pattern of stars.

corona A layer of gas around the Sun, which can be seen during an eclipse of the Sun.

galaxy A group of millions of stars.

light year The distance light travels in one year.

mass The amount of matter in anything. An object's weight is affected by gravity; its mass is always the same.

nebula A cloud of gas and dust. The plural of nebula is nebulae.

northern hemisphere The top half of the Earth.

orbit The continuous journey of one object around another.

rotate To spin around.

solar system The Sun and the planets, moons and asteroids which orbit it.

southern hemisphere The bottom half of the Earth.

spectrum The colours which make up light.

star A large ball of very hot gas which gives off light as it rotates in space.

Index

Photographic credits: Bruce Coleman Ltd (E Crichton) 3, (S Nielsen) 26; Eye Ubiquitous (F Leather) 7; Robert Harding (R Francis) 19; Science Photo Library (NASA) 8, (R Royer) 11, 13, (J Sanford) 15, (NOAO) 20, 25, (D Di Cicco) 22; ZEFA 5, 17.